To the Beach

Dona Herweck Rice

Publishing Credits

Rachelle Cracchiolo, M.S.Ed., *Publisher*
Conni Medina, M.A.Ed., *Managing Editor*
Nika Fabienke, Ed.D., *Content Director*
Véronique Bos, *Creative Director*
Shaun N. Bernadou, *Art Director*
Valerie Morales, *Associate Editor*
John Leach, *Assistant Editor*
Courtney Roberson, *Senior Graphic Designer*

Image Credits: All images from iStock and/or Shutterstock.

Teacher Created Materials
5301 Oceanus Drive
Huntington Beach, CA 92649-1030
www.tcmpub.com

ISBN 978-1-4938-9837-4
© 2019 Teacher Created Materials, Inc.
Printed by: 590835
Printed in: Taiwan

I0896092

When we go to the

beach

, we will

park

.

When we go to the

 , we will  .

beach run

When we go to the

beach

, we will

splash

.

When we go to the

beach
, we will

swim
.

When we go to the

beach
, we will

dig
.

When we go to the

beach

, we will

take pictures

.

When we go to the

 , we will  .

beach eat

When we go to the

, we will .

beach drink

When we go to the

beach

, we will

float

.

When we go to the

beach

, we will

surf

.

High-Frequency Words

New Words

go we
when will

Review Words

the to